THIS BOOK BELONGS TO:

CONTACT INFORMATION	
NAME:	
ADDRESS:	
PHONE:	

START / END DATES

_____ / _____ / _____   TO   _____ / _____ / _____

DEDICATION

This book is dedicated to all the amazing goat owners who love raising and taking care of GOATS!

You are my inspiration in producing books and I'm excited to help in the planning of GOAT CARE into your day around the world!

How to Use this Goat Record Keeping Notebook:

The purpose of this Goat Raising Log Book is for anyone is to keep all various goat breeding activities and information organized in one easy to find spot.

Here are some simple guidelines to follow so you can make the most of using this book:

1. The first "Goat Information" section is for you to write out your goat's name, tattoo, breed, physical characteristics and a pedigree chart so you can track your goat raising adventures.

2. Most ideas are inspired by something we have seen. Use the "Medical Information" section to write down the date, nature of any illness, parasite control, testing record and a vaccination record so you can go back there to be reminded later.

3. The "Doe's Kidding Record" section is for you to write out THAT Doe's name, date, breed, kidding date and any important information for each Doe.

4. Some ideas require listing them out, the "Buck;s Record Of Progeny" section is great for using this to record the year, bred to, kids and much more for you to use to keep track and refer back to this list later.

5. The "Goat Record" section is so you can list out the Goat's name, breed, Identification, date of birth, weaning date, and month to month weight tracker... and be inspired to add to your goat record.

6. And finally pages with a "Milk Production" section for you to make entries about month to month milk produced, yearly production and the value per lbs... and much much more.

Have fun!

GOAT INFORMATION

PHOTO

NAME	☐ BUCK	☐ DOE
BREED	BIRTH DATE:	

DATE ACQUIRED:	HOW ACQUIRED: ☐ BORN ON FARM ☐ PURCHASED ☐ LEASED

COLORS / IDENTIFYING MARKS:

PURPOSE:	☐ MILK	☐ MEAT	☐ PET	☐ OTHER

PEDIGREE CHART

- SIRE
  - GRAND SIRE
    - GREAT GRAND SIRE
    - GREAT GRAND DAM
  - GRAND DAM
    - GREAT GRAND SIRE
    - GREAT GRAND DAM
- DAM
  - GRAND SIRE
    - GREAT GRAND SIRE
    - GREAT GRAND DAM
  - GRAND DAM
    - GREAT GRAND SIRE
    - GREAT GRAND DAM

MEDICAL INFORMATION

INJURY OR ILLNESS

DATE	DESCRIPTION OR NATURE OF ILLNESS	TREATMENT

PARASITE CONTROL

DATE	METHOD OR DEWORMER	DATE	METHOD OR DEWORMER

TESTING RECORD

DATE	TEST PERFORMED (CAE, CL, TB...)	RESULT	DATE	TEST PERFORMED (CAE, CL, TB...)	RESULT

INJURY OR ILLNESS

DATE	TARGET DISEASE	DRUG OR SUPPLEMENT USED	DOSAGE	RESULTS

DOE'S KIDDING RECORD

DOE'S NAME:

DATE BREED	KIDDING DATE	# OF KIDS	SEX D/B	NAME OF KID	SIRE OF KID	WEIGHT	TATTOO

BUCK'S RECORD OF PROGENY

DOE'S NAME:	

YEAR	BRED TO	KIDS	DOE/BUCK

GOAT RECORD

GOAT'S NAME:		IDENTIFICATION:	
BREED:	DATE OF BIRTH:		DATE OF WEANED:

WEIGHT (POUNDS)

BIRTH	JAN	FEB	MAR	APR	MAY	JUN	JUL	AUG	SEPT	OCT	NOV	DEC	FINAL

FEED RECORD

	JAN	FEB	MAR	APR	MAY	JUN	JUL	AUG	SEPT	OCT	NOV	DEC	TOTAL
GRAIN													
GRAIN													
PASTURE													

MILK PRODUCTION

GOAT'S NAME:		IDENTIFICATION:	
BREED:	DATE OF BIRTH:	KIDDING DATE:	

JANUARY		AVERAGE LBS / DAY X 31 DAYS =		LBS
FEBRUARY		AVERAGE LBS / DAY X 31 DAYS =		LBS
MARCH		AVERAGE LBS / DAY X 31 DAYS =		LBS
APRIL		AVERAGE LBS / DAY X 31 DAYS =		LBS
MAY		AVERAGE LBS / DAY X 31 DAYS =		LBS
JUNE		AVERAGE LBS / DAY X 31 DAYS =		LBS
JULY		AVERAGE LBS / DAY X 31 DAYS =		LBS
AUGUST		AVERAGE LBS / DAY X 31 DAYS =		LBS
SEPTEMBER		AVERAGE LBS / DAY X 31 DAYS =		LBS
OCTOBER		AVERAGE LBS / DAY X 31 DAYS =		LBS
NOVEMBER		AVERAGE LBS / DAY X 31 DAYS =		LBS
DECEMBER		AVERAGE LBS / DAY X 31 DAYS =		LBS
YEARLY TOTAL MILK PRODUCED =				LBS

TOTAL VALUE OF MILK PRODUCED FOR THE YEAR

	LBS X $		VALUE PER LBS =	

GOAT INFORMATION

PHOTO

NAME	☐ BUCK	☐ DOE
BREED	BIRTH DATE:	

DATE ACQUIRED:	HOW ACQUIRED: ☐ BORN ON FARM ☐ PURCHASED ☐ LEASED

COLORS / IDENTIFYING MARKS:

PURPOSE:	☐ MILK	☐ MEAT	☐ PET	☐ OTHER

PEDIGREE CHART

- SIRE
  - GRAND SIRE
    - GREAT GRAND SIRE
    - GREAT GRAND DAM
  - GRAND DAM
    - GREAT GRAND SIRE
    - GREAT GRAND DAM
- DAM
  - GRAND SIRE
    - GREAT GRAND SIRE
    - GREAT GRAND DAM
  - GRAND DAM
    - GREAT GRAND SIRE
    - GREAT GRAND DAM

MEDICAL INFORMATION

INJURY OR ILLNESS

DATE	DESCRIPTION OR NATURE OF ILLNESS	TREATMENT

PARASITE CONTROL

DATE	METHOD OR DEWORMER	DATE	METHOD OR DEWORMER

TESTING RECORD

DATE	TEST PERFORMED (CAE, CL, TB...)	RESULT	DATE	TEST PERFORMED (CAE, CL, TB...)	RESULT

INJURY OR ILLNESS

DATE	TARGET DISEASE	DRUG OR SUPPLEMENT USED	DOSAGE	RESULTS

DOE'S KIDDING RECORD

DOE'S NAME:

DATE BREED	KIDDING DATE	# OF KIDS	SEX D/B	NAME OF KID	SIRE OF KID	WEIGHT	TATTOO

BUCK'S RECORD OF PROGENY

DOE'S NAME:	

YEAR	BRED TO	KIDS	DOE/BUCK

GOAT RECORD

GOAT'S NAME:		IDENTIFICATION:	
BREED:	DATE OF BIRTH:		DATE OF WEANED:

WEIGHT (POUNDS)

BIRTH	JAN	FEB	MAR	APR	MAY	JUN	JUL	AUG	SEPT	OCT	NOV	DEC	FINAL

FEED RECORD

	JAN	FEB	MAR	APR	MAY	JUN	JUL	AUG	SEPT	OCT	NOV	DEC	TOTAL
GRAIN													
GRAIN													
PASTURE													

MILK PRODUCTION

GOAT'S NAME:	IDENTIFICATION:
BREED:	DATE OF BIRTH:	KIDDING DATE:

JANUARY		AVERAGE LBS / DAY X 31 DAYS =		LBS
FEBRUARY		AVERAGE LBS / DAY X 31 DAYS =		LBS
MARCH		AVERAGE LBS / DAY X 31 DAYS =		LBS
APRIL		AVERAGE LBS / DAY X 31 DAYS =		LBS
MAY		AVERAGE LBS / DAY X 31 DAYS =		LBS
JUNE		AVERAGE LBS / DAY X 31 DAYS =		LBS
JULY		AVERAGE LBS / DAY X 31 DAYS =		LBS
AUGUST		AVERAGE LBS / DAY X 31 DAYS =		LBS
SEPTEMBER		AVERAGE LBS / DAY X 31 DAYS =		LBS
OCTOBER		AVERAGE LBS / DAY X 31 DAYS =		LBS
NOVEMBER		AVERAGE LBS / DAY X 31 DAYS =		LBS
DECEMBER		AVERAGE LBS / DAY X 31 DAYS =		LBS
YEARLY TOTAL MILK PRODUCED =				LBS

TOTAL VALUE OF MILK PRODUCED FOR THE YEAR

	LBS X $		VALUE PER LBS =	

GOAT INFORMATION

PHOTO

NAME		☐ BUCK	☐ DOE
BREED		BIRTH DATE:	
DATE ACQUIRED:	HOW ACQUIRED: ☐ BORN ON FARM ☐ PURCHASED ☐ LEASED		
COLORS / IDENTIFYING MARKS:			
PURPOSE: ☐ MILK ☐ MEAT ☐ PET	☐ OTHER		

PEDIGREE CHART

- SIRE
  - GRAND SIRE
    - GREAT GRAND SIRE
    - GREAT GRAND DAM
  - GRAND DAM
    - GREAT GRAND SIRE
    - GREAT GRAND DAM
- DAM
  - GRAND SIRE
    - GREAT GRAND SIRE
    - GREAT GRAND DAM
  - GRAND DAM
    - GREAT GRAND SIRE
    - GREAT GRAND DAM

MEDICAL INFORMATION

INJURY OR ILLNESS

DATE	DESCRIPTION OR NATURE OF ILLNESS	TREATMENT

PARASITE CONTROL

DATE	METHOD OR DEWORMER		DATE	METHOD OR DEWORMER

TESTING RECORD

DATE	TEST PERFORMED (CAE, CL, TB...)	RESULT		DATE	TEST PERFORMED (CAE, CL, TB...)	RESULT

INJURY OR ILLNESS

DATE	TARGET DISEASE	DRUG OR SUPPLEMENT USED	DOSAGE	RESULTS

DOE'S KIDDING RECORD

DOE'S NAME:

DATE BREED	KIDDING DATE	# OF KIDS	SEX D/B	NAME OF KID	SIRE OF KID	WEIGHT	TATTOO

BUCK'S RECORD OF PROGENY

DOE'S NAME:

YEAR	BRED TO	KIDS	DOE/BUCK

GOAT RECORD

GOAT'S NAME:		IDENTIFICATION:	
BREED:	DATE OF BIRTH:		DATE OF WEANED:

WEIGHT (POUNDS)

BIRTH	JAN	FEB	MAR	APR	MAY	JUN	JUL	AUG	SEPT	OCT	NOV	DEC	FINAL

FEED RECORD

	JAN	FEB	MAR	APR	MAY	JUN	JUL	AUG	SEPT	OCT	NOV	DEC	TOTAL
GRAIN													
GRAIN													
PASTURE													

MILK PRODUCTION

GOAT'S NAME:	IDENTIFICATION:

BREED:	DATE OF BIRTH:	KIDDING DATE:

JANUARY		AVERAGE LBS / DAY X 31 DAYS =		LBS
FEBRUARY		AVERAGE LBS / DAY X 31 DAYS =		LBS
MARCH		AVERAGE LBS / DAY X 31 DAYS =		LBS
APRIL		AVERAGE LBS / DAY X 31 DAYS =		LBS
MAY		AVERAGE LBS / DAY X 31 DAYS =		LBS
JUNE		AVERAGE LBS / DAY X 31 DAYS =		LBS
JULY		AVERAGE LBS / DAY X 31 DAYS =		LBS
AUGUST		AVERAGE LBS / DAY X 31 DAYS =		LBS
SEPTEMBER		AVERAGE LBS / DAY X 31 DAYS =		LBS
OCTOBER		AVERAGE LBS / DAY X 31 DAYS =		LBS
NOVEMBER		AVERAGE LBS / DAY X 31 DAYS =		LBS
DECEMBER		AVERAGE LBS / DAY X 31 DAYS =		LBS
YEARLY TOTAL MILK PRODUCED =				LBS

TOTAL VALUE OF MILK PRODUCED FOR THE YEAR

	LBS X $		VALUE PER LBS =	

GOAT INFORMATION

PHOTO

NAME	☐ BUCK	☐ DOE
BREED	BIRTH DATE:	
DATE ACQUIRED:	HOW ACQUIRED: ☐ BORN ON FARM ☐ PURCHASED ☐ LEASED	
COLORS / IDENTIFYING MARKS:		
PURPOSE: ☐ MILK ☐ MEAT ☐ PET ☐ OTHER		

PEDIGREE CHART

- SIRE
  - GRAND SIRE
    - GREAT GRAND SIRE
    - GREAT GRAND DAM
  - GRAND DAM
    - GREAT GRAND SIRE
    - GREAT GRAND DAM
- DAM
  - GRAND SIRE
    - GREAT GRAND SIRE
    - GREAT GRAND DAM
  - GRAND DAM
    - GREAT GRAND SIRE
    - GREAT GRAND DAM

MEDICAL INFORMATION

INJURY OR ILLNESS

DATE	DESCRIPTION OR NATURE OF ILLNESS	TREATMENT

PARASITE CONTROL

DATE	METHOD OR DEWORMER	DATE	METHOD OR DEWORMER

TESTING RECORD

DATE	TEST PERFORMED (CAE, CL, TB...)	RESULT	DATE	TEST PERFORMED (CAE, CL, TB...)	RESULT

INJURY OR ILLNESS

DATE	TARGET DISEASE	DRUG OR SUPPLEMENT USED	DOSAGE	RESULTS

DOE'S KIDDING RECORD

DOE'S NAME:

DATE BREED	KIDDING DATE	# OF KIDS	SEX D/B	NAME OF KID	SIRE OF KID	WEIGHT	TATTOO

BUCK'S RECORD OF PROGENY

DOE'S NAME:	

YEAR	BRED TO	KIDS	DOE/BUCK

GOAT RECORD

GOAT'S NAME:		IDENTIFICATION:	
BREED:	DATE OF BIRTH:		DATE OF WEANED:

WEIGHT (POUNDS)

BIRTH	JAN	FEB	MAR	APR	MAY	JUN	JUL	AUG	SEPT	OCT	NOV	DEC	FINAL

FEED RECORD

	JAN	FEB	MAR	APR	MAY	JUN	JUL	AUG	SEPT	OCT	NOV	DEC	TOTAL
GRAIN													
GRAIN													
PASTURE													

MILK PRODUCTION

GOAT'S NAME:		IDENTIFICATION:		
BREED:		DATE OF BIRTH:	KIDDING DATE:	

JANUARY		AVERAGE LBS / DAY X 31 DAYS =		LBS
FEBRUARY		AVERAGE LBS / DAY X 31 DAYS =		LBS
MARCH		AVERAGE LBS / DAY X 31 DAYS =		LBS
APRIL		AVERAGE LBS / DAY X 31 DAYS =		LBS
MAY		AVERAGE LBS / DAY X 31 DAYS =		LBS
JUNE		AVERAGE LBS / DAY X 31 DAYS =		LBS
JULY		AVERAGE LBS / DAY X 31 DAYS =		LBS
AUGUST		AVERAGE LBS / DAY X 31 DAYS =		LBS
SEPTEMBER		AVERAGE LBS / DAY X 31 DAYS =		LBS
OCTOBER		AVERAGE LBS / DAY X 31 DAYS =		LBS
NOVEMBER		AVERAGE LBS / DAY X 31 DAYS =		LBS
DECEMBER		AVERAGE LBS / DAY X 31 DAYS =		LBS
YEARLY TOTAL MILK PRODUCED =				LBS

TOTAL VALUE OF MILK PRODUCED FOR THE YEAR

	LBS X $		VALUE PER LBS =	

GOAT INFORMATION

PHOTO

NAME		☐ BUCK	☐ DOE
BREED		BIRTH DATE:	
DATE ACQUIRED:	HOW ACQUIRED: ☐ BORN ON FARM ☐ PURCHASED ☐ LEASED		
COLORS / IDENTIFYING MARKS:			

PURPOSE:	☐ MILK	☐ MEAT	☐ PET	☐ OTHER

PEDIGREE CHART

- SIRE
  - GRAND SIRE
    - GREAT GRAND SIRE
    - GREAT GRAND DAM
  - GRAND DAM
    - GREAT GRAND SIRE
    - GREAT GRAND DAM
- DAM
  - GRAND SIRE
    - GREAT GRAND SIRE
    - GREAT GRAND DAM
  - GRAND DAM
    - GREAT GRAND SIRE
    - GREAT GRAND DAM

MEDICAL INFORMATION

INJURY OR ILLNESS

DATE	DESCRIPTION OR NATURE OF ILLNESS	TREATMENT

PARASITE CONTROL

DATE	METHOD OR DEWORMER	DATE	METHOD OR DEWORMER

TESTING RECORD

DATE	TEST PERFORMED (CAE, CL, TB...)	RESULT	DATE	TEST PERFORMED (CAE, CL, TB...)	RESULT

INJURY OR ILLNESS

DATE	TARGET DISEASE	DRUG OR SUPPLEMENT USED	DOSAGE	RESULTS

DOE'S KIDDING RECORD

DOE'S NAME:

DATE BREED	KIDDING DATE	# OF KIDS	SEX D/B	NAME OF KID	SIRE OF KID	WEIGHT	TATTOO

BUCK'S RECORD OF PROGENY

DOE'S NAME:

YEAR	BRED TO	KIDS	DOE/BUCK

GOAT RECORD

GOAT'S NAME:		IDENTIFICATION:	
BREED:	DATE OF BIRTH:		DATE OF WEANED:

WEIGHT (POUNDS)

BIRTH	JAN	FEB	MAR	APR	MAY	JUN	JUL	AUG	SEPT	OCT	NOV	DEC	FINAL

FEED RECORD

	JAN	FEB	MAR	APR	MAY	JUN	JUL	AUG	SEPT	OCT	NOV	DEC	TOTAL
GRAIN													
GRAIN													
PASTURE													

MILK PRODUCTION

GOAT'S NAME:		IDENTIFICATION:		

BREED:		DATE OF BIRTH:		KIDDING DATE:	

JANUARY		AVERAGE LBS / DAY X 31 DAYS =		LBS
FEBRUARY		AVERAGE LBS / DAY X 31 DAYS =		LBS
MARCH		AVERAGE LBS / DAY X 31 DAYS =		LBS
APRIL		AVERAGE LBS / DAY X 31 DAYS =		LBS
MAY		AVERAGE LBS / DAY X 31 DAYS =		LBS
JUNE		AVERAGE LBS / DAY X 31 DAYS =		LBS
JULY		AVERAGE LBS / DAY X 31 DAYS =		LBS
AUGUST		AVERAGE LBS / DAY X 31 DAYS =		LBS
SEPTEMBER		AVERAGE LBS / DAY X 31 DAYS =		LBS
OCTOBER		AVERAGE LBS / DAY X 31 DAYS =		LBS
NOVEMBER		AVERAGE LBS / DAY X 31 DAYS =		LBS
DECEMBER		AVERAGE LBS / DAY X 31 DAYS =		LBS
YEARLY TOTAL MILK PRODUCED =				LBS

TOTAL VALUE OF MILK PRODUCED FOR THE YEAR

	LBS X $		VALUE PER LBS =	

GOAT INFORMATION

PHOTO

NAME		☐ BUCK	☐ DOE
BREED		BIRTH DATE:	
DATE ACQUIRED:	HOW ACQUIRED: ☐ BORN ON FARM ☐ PURCHASED ☐ LEASED		
COLORS / IDENTIFYING MARKS:			
PURPOSE: ☐ MILK ☐ MEAT ☐ PET ☐ OTHER			

PEDIGREE CHART

SIRE

GRAND SIRE

GREAT GRAND SIRE

GREAT GRAND DAM

GRAND DAM

GREAT GRAND SIRE

GREAT GRAND DAM

DAM

GRAND SIRE

GREAT GRAND SIRE

GREAT GRAND DAM

GRAND DAM

GREAT GRAND SIRE

GREAT GRAND DAM

MEDICAL INFORMATION

INJURY OR ILLNESS

DATE	DESCRIPTION OR NATURE OF ILLNESS	TREATMENT

PARASITE CONTROL

DATE	METHOD OR DEWORMER	DATE	METHOD OR DEWORMER

TESTING RECORD

DATE	TEST PERFORMED (CAE, CL, TB...)	RESULT	DATE	TEST PERFORMED (CAE, CL, TB...)	RESULT

INJURY OR ILLNESS

DATE	TARGET DISEASE	DRUG OR SUPPLEMENT USED	DOSAGE	RESULTS

DOE'S KIDDING RECORD

DOE'S NAME:	

DATE BREED	KIDDING DATE	# OF KIDS	SEX D/B	NAME OF KID	SIRE OF KID	WEIGHT	TATTOO

BUCK'S RECORD OF PROGENY

DOE'S NAME:	

YEAR	BRED TO	KIDS	DOE/BUCK

GOAT RECORD

GOAT'S NAME:

IDENTIFICATION:

BREED:

DATE OF BIRTH:

DATE OF WEANED:

WEIGHT (POUNDS)

BIRTH	JAN	FEB	MAR	APR	MAY	JUN	JUL	AUG	SEPT	OCT	NOV	DEC	FINAL

FEED RECORD

	JAN	FEB	MAR	APR	MAY	JUN	JUL	AUG	SEPT	OCT	NOV	DEC	TOTAL
GRAIN													
GRAIN													
PASTURE													

MILK PRODUCTION

GOAT'S NAME:		IDENTIFICATION:	
BREED:	DATE OF BIRTH:	KIDDING DATE:	

JANUARY		AVERAGE LBS / DAY X 31 DAYS =		LBS
FEBRUARY		AVERAGE LBS / DAY X 31 DAYS =		LBS
MARCH		AVERAGE LBS / DAY X 31 DAYS =		LBS
APRIL		AVERAGE LBS / DAY X 31 DAYS =		LBS
MAY		AVERAGE LBS / DAY X 31 DAYS =		LBS
JUNE		AVERAGE LBS / DAY X 31 DAYS =		LBS
JULY		AVERAGE LBS / DAY X 31 DAYS =		LBS
AUGUST		AVERAGE LBS / DAY X 31 DAYS =		LBS
SEPTEMBER		AVERAGE LBS / DAY X 31 DAYS =		LBS
OCTOBER		AVERAGE LBS / DAY X 31 DAYS =		LBS
NOVEMBER		AVERAGE LBS / DAY X 31 DAYS =		LBS
DECEMBER		AVERAGE LBS / DAY X 31 DAYS =		LBS
YEARLY TOTAL MILK PRODUCED =				LBS

TOTAL VALUE OF MILK PRODUCED FOR THE YEAR

	LBS X $		VALUE PER LBS =	

GOAT INFORMATION

PHOTO

NAME	☐ BUCK	☐ DOE
BREED	BIRTH DATE:	

DATE ACQUIRED:	HOW ACQUIRED: ☐ BORN ON FARM ☐ PURCHASED ☐ LEASED

COLORS / IDENTIFYING MARKS:

PURPOSE:	☐ MILK	☐ MEAT	☐ PET	☐ OTHER

PEDIGREE CHART

SIRE

GRAND SIRE

GREAT GRAND SIRE

GREAT GRAND DAM

GRAND DAM

GREAT GRAND SIRE

GREAT GRAND DAM

DAM

GRAND SIRE

GREAT GRAND SIRE

GREAT GRAND DAM

GRAND DAM

GREAT GRAND SIRE

GREAT GRAND DAM

MEDICAL INFORMATION

INJURY OR ILLNESS

DATE	DESCRIPTION OR NATURE OF ILLNESS	TREATMENT

PARASITE CONTROL

DATE	METHOD OR DEWORMER	DATE	METHOD OR DEWORMER

TESTING RECORD

DATE	TEST PERFORMED (CAE, CL, TB...)	RESULT	DATE	TEST PERFORMED (CAE, CL, TB...)	RESULT

INJURY OR ILLNESS

DATE	TARGET DISEASE	DRUG OR SUPPLEMENT USED	DOSAGE	RESULTS

DOE'S KIDDING RECORD

DOE'S NAME:

DATE BREED	KIDDING DATE	# OF KIDS	SEX D/B	NAME OF KID	SIRE OF KID	WEIGHT	TATTOO

BUCK'S RECORD OF PROGENY

DOE'S NAME:	

YEAR	BRED TO	KIDS	DOE/BUCK

GOAT RECORD

GOAT'S NAME:		IDENTIFICATION:	
BREED:	DATE OF BIRTH:	DATE OF WEANED:	

WEIGHT (POUNDS)

BIRTH	JAN	FEB	MAR	APR	MAY	JUN	JUL	AUG	SEPT	OCT	NOV	DEC	FINAL

FEED RECORD

	JAN	FEB	MAR	APR	MAY	JUN	JUL	AUG	SEPT	OCT	NOV	DEC	TOTAL
GRAIN													
GRAIN													
PASTURE													

MILK PRODUCTION

GOAT'S NAME:	IDENTIFICATION:	
BREED:	DATE OF BIRTH:	KIDDING DATE:

JANUARY		AVERAGE LBS / DAY X 31 DAYS =		LBS
FEBRUARY		AVERAGE LBS / DAY X 31 DAYS =		LBS
MARCH		AVERAGE LBS / DAY X 31 DAYS =		LBS
APRIL		AVERAGE LBS / DAY X 31 DAYS =		LBS
MAY		AVERAGE LBS / DAY X 31 DAYS =		LBS
JUNE		AVERAGE LBS / DAY X 31 DAYS =		LBS
JULY		AVERAGE LBS / DAY X 31 DAYS =		LBS
AUGUST		AVERAGE LBS / DAY X 31 DAYS =		LBS
SEPTEMBER		AVERAGE LBS / DAY X 31 DAYS =		LBS
OCTOBER		AVERAGE LBS / DAY X 31 DAYS =		LBS
NOVEMBER		AVERAGE LBS / DAY X 31 DAYS =		LBS
DECEMBER		AVERAGE LBS / DAY X 31 DAYS =		LBS
YEARLY TOTAL MILK PRODUCED =				LBS

TOTAL VALUE OF MILK PRODUCED FOR THE YEAR

	LBS X $		VALUE PER LBS =	

GOAT INFORMATION

PHOTO

NAME	☐ BUCK	☐ DOE
BREED	BIRTH DATE:	
DATE ACQUIRED:	HOW ACQUIRED: ☐ BORN ON FARM ☐ PURCHASED ☐ LEASED	
COLORS / IDENTIFYING MARKS:		
PURPOSE: ☐ MILK ☐ MEAT ☐ PET ☐ OTHER		

PEDIGREE CHART

SIRE

GRAND SIRE
- GREAT GRAND SIRE
- GREAT GRAND DAM

GRAND DAM
- GREAT GRAND SIRE
- GREAT GRAND DAM

DAM

GRAND SIRE
- GREAT GRAND SIRE
- GREAT GRAND DAM

GRAND DAM
- GREAT GRAND SIRE
- GREAT GRAND DAM

MEDICAL INFORMATION

INJURY OR ILLNESS

DATE	DESCRIPTION OR NATURE OF ILLNESS	TREATMENT

PARASITE CONTROL

DATE	METHOD OR DEWORMER	DATE	METHOD OR DEWORMER

TESTING RECORD

DATE	TEST PERFORMED (CAE, CL, TB...)	RESULT	DATE	TEST PERFORMED (CAE, CL, TB...)	RESULT

INJURY OR ILLNESS

DATE	TARGET DISEASE	DRUG OR SUPPLEMENT USED	DOSAGE	RESULTS

DOE'S KIDDING RECORD

DOE'S NAME:

DATE BREED	KIDDING DATE	# OF KIDS	SEX D/B	NAME OF KID	SIRE OF KID	WEIGHT	TATTOO

BUCK'S RECORD OF PROGENY

DOE'S NAME:

YEAR	BRED TO	KIDS	DOE/BUCK

GOAT RECORD

GOAT'S NAME:		IDENTIFICATION:	
BREED:	DATE OF BIRTH:		DATE OF WEANED:

WEIGHT (POUNDS)

BIRTH	JAN	FEB	MAR	APR	MAY	JUN	JUL	AUG	SEPT	OCT	NOV	DEC	FINAL

FEED RECORD

	JAN	FEB	MAR	APR	MAY	JUN	JUL	AUG	SEPT	OCT	NOV	DEC	TOTAL
GRAIN													
GRAIN													
PASTURE													

MILK PRODUCTION

GOAT'S NAME:		IDENTIFICATION:	
BREED:	DATE OF BIRTH:	KIDDING DATE:	

JANUARY		AVERAGE LBS / DAY X 31 DAYS =		LBS
FEBRUARY		AVERAGE LBS / DAY X 31 DAYS =		LBS
MARCH		AVERAGE LBS / DAY X 31 DAYS =		LBS
APRIL		AVERAGE LBS / DAY X 31 DAYS =		LBS
MAY		AVERAGE LBS / DAY X 31 DAYS =		LBS
JUNE		AVERAGE LBS / DAY X 31 DAYS =		LBS
JULY		AVERAGE LBS / DAY X 31 DAYS =		LBS
AUGUST		AVERAGE LBS / DAY X 31 DAYS =		LBS
SEPTEMBER		AVERAGE LBS / DAY X 31 DAYS =		LBS
OCTOBER		AVERAGE LBS / DAY X 31 DAYS =		LBS
NOVEMBER		AVERAGE LBS / DAY X 31 DAYS =		LBS
DECEMBER		AVERAGE LBS / DAY X 31 DAYS =		LBS
YEARLY TOTAL MILK PRODUCED =				LBS

TOTAL VALUE OF MILK PRODUCED FOR THE YEAR

	LBS X $		VALUE PER LBS =	

GOAT INFORMATION

PHOTO

NAME	☐ BUCK	☐ DOE
BREED	BIRTH DATE:	
DATE ACQUIRED:	HOW ACQUIRED: ☐ BORN ON FARM ☐ PURCHASED ☐ LEASED	
COLORS / IDENTIFYING MARKS:		
PURPOSE: ☐ MILK ☐ MEAT ☐ PET ☐ OTHER		

PEDIGREE CHART

- SIRE
  - GRAND SIRE
    - GREAT GRAND SIRE
    - GREAT GRAND DAM
  - GRAND DAM
    - GREAT GRAND SIRE
    - GREAT GRAND DAM
- DAM
  - GRAND SIRE
    - GREAT GRAND SIRE
    - GREAT GRAND DAM
  - GRAND DAM
    - GREAT GRAND SIRE
    - GREAT GRAND DAM

MEDICAL INFORMATION

INJURY OR ILLNESS

DATE	DESCRIPTION OR NATURE OF ILLNESS	TREATMENT

PARASITE CONTROL

DATE	METHOD OR DEWORMER	DATE	METHOD OR DEWORMER

TESTING RECORD

DATE	TEST PERFORMED (CAE, CL, TB...)	RESULT	DATE	TEST PERFORMED (CAE, CL, TB...)	RESULT

INJURY OR ILLNESS

DATE	TARGET DISEASE	DRUG OR SUPPLEMENT USED	DOSAGE	RESULTS

DOE'S KIDDING RECORD

DOE'S NAME:	

DATE BREED	KIDDING DATE	# OF KIDS	SEX D/B	NAME OF KID	SIRE OF KID	WEIGHT	TATTOO

BUCK'S RECORD OF PROGENY

DOE'S NAME:	

YEAR	BRED TO	KIDS	DOE/BUCK

GOAT RECORD

GOAT'S NAME:		IDENTIFICATION:	
BREED:	DATE OF BIRTH:		DATE OF WEANED:

WEIGHT (POUNDS)

BIRTH	JAN	FEB	MAR	APR	MAY	JUN	JUL	AUG	SEPT	OCT	NOV	DEC	FINAL

FEED RECORD

	JAN	FEB	MAR	APR	MAY	JUN	JUL	AUG	SEPT	OCT	NOV	DEC	TOTAL
GRAIN													
GRAIN													
PASTURE													

MILK PRODUCTION

GOAT'S NAME:		IDENTIFICATION:	
BREED:	DATE OF BIRTH:	KIDDING DATE:	

JANUARY		AVERAGE LBS / DAY X 31 DAYS =		LBS
FEBRUARY		AVERAGE LBS / DAY X 31 DAYS =		LBS
MARCH		AVERAGE LBS / DAY X 31 DAYS =		LBS
APRIL		AVERAGE LBS / DAY X 31 DAYS =		LBS
MAY		AVERAGE LBS / DAY X 31 DAYS =		LBS
JUNE		AVERAGE LBS / DAY X 31 DAYS =		LBS
JULY		AVERAGE LBS / DAY X 31 DAYS =		LBS
AUGUST		AVERAGE LBS / DAY X 31 DAYS =		LBS
SEPTEMBER		AVERAGE LBS / DAY X 31 DAYS =		LBS
OCTOBER		AVERAGE LBS / DAY X 31 DAYS =		LBS
NOVEMBER		AVERAGE LBS / DAY X 31 DAYS =		LBS
DECEMBER		AVERAGE LBS / DAY X 31 DAYS =		LBS
YEARLY TOTAL MILK PRODUCED =				LBS

TOTAL VALUE OF MILK PRODUCED FOR THE YEAR

	LBS X $		VALUE PER LBS =	

GOAT INFORMATION

PHOTO

NAME	☐ BUCK	☐ DOE
BREED	BIRTH DATE:	
DATE ACQUIRED:	HOW ACQUIRED: ☐ BORN ON FARM ☐ PURCHASED ☐ LEASED	
COLORS / IDENTIFYING MARKS:		
PURPOSE: ☐ MILK ☐ MEAT ☐ PET ☐ OTHER		

PEDIGREE CHART

SIRE

GRAND SIRE

GREAT GRAND SIRE

GREAT GRAND DAM

GRAND DAM

GREAT GRAND SIRE

GREAT GRAND DAM

DAM

GRAND SIRE

GREAT GRAND SIRE

GREAT GRAND DAM

GRAND DAM

GREAT GRAND SIRE

GREAT GRAND DAM

MEDICAL INFORMATION

INJURY OR ILLNESS

DATE	DESCRIPTION OR NATURE OF ILLNESS	TREATMENT

PARASITE CONTROL

DATE	METHOD OR DEWORMER	DATE	METHOD OR DEWORMER

TESTING RECORD

DATE	TEST PERFORMED (CAE, CL, TB...)	RESULT	DATE	TEST PERFORMED (CAE, CL, TB...)	RESULT

INJURY OR ILLNESS

DATE	TARGET DISEASE	DRUG OR SUPPLEMENT USED	DOSAGE	RESULTS

DOE'S KIDDING RECORD

DOE'S NAME:	

DATE BREED	KIDDING DATE	# OF KIDS	SEX D/B	NAME OF KID	SIRE OF KID	WEIGHT	TATTOO

BUCK'S RECORD OF PROGENY

DOE'S NAME:	

YEAR	BRED TO	KIDS	DOE/BUCK

GOAT RECORD

GOAT'S NAME: IDENTIFICATION:

BREED: DATE OF BIRTH: DATE OF WEANED:

WEIGHT (POUNDS)

BIRTH	JAN	FEB	MAR	APR	MAY	JUN	JUL	AUG	SEPT	OCT	NOV	DEC	FINAL

FEED RECORD

	JAN	FEB	MAR	APR	MAY	JUN	JUL	AUG	SEPT	OCT	NOV	DEC	TOTAL
GRAIN													
GRAIN													
PASTURE													

MILK PRODUCTION

GOAT'S NAME:		IDENTIFICATION:	
BREED:	DATE OF BIRTH:	KIDDING DATE:	

JANUARY		AVERAGE LBS / DAY X 31 DAYS =		LBS
FEBRUARY		AVERAGE LBS / DAY X 31 DAYS =		LBS
MARCH		AVERAGE LBS / DAY X 31 DAYS =		LBS
APRIL		AVERAGE LBS / DAY X 31 DAYS =		LBS
MAY		AVERAGE LBS / DAY X 31 DAYS =		LBS
JUNE		AVERAGE LBS / DAY X 31 DAYS =		LBS
JULY		AVERAGE LBS / DAY X 31 DAYS =		LBS
AUGUST		AVERAGE LBS / DAY X 31 DAYS =		LBS
SEPTEMBER		AVERAGE LBS / DAY X 31 DAYS =		LBS
OCTOBER		AVERAGE LBS / DAY X 31 DAYS =		LBS
NOVEMBER		AVERAGE LBS / DAY X 31 DAYS =		LBS
DECEMBER		AVERAGE LBS / DAY X 31 DAYS =		LBS
YEARLY TOTAL MILK PRODUCED =				LBS

TOTAL VALUE OF MILK PRODUCED FOR THE YEAR

	LBS X $		VALUE PER LBS =	

GOAT INFORMATION

PHOTO

NAME	☐ BUCK	☐ DOE
BREED	BIRTH DATE:	
DATE ACQUIRED:	HOW ACQUIRED: ☐ BORN ON FARM ☐ PURCHASED ☐ LEASED	
COLORS / IDENTIFYING MARKS:		
PURPOSE: ☐ MILK ☐ MEAT ☐ PET ☐ OTHER		

PEDIGREE CHART

SIRE

GRAND SIRE

GREAT GRAND SIRE

GREAT GRAND DAM

GRAND DAM

GREAT GRAND SIRE

GREAT GRAND DAM

DAM

GRAND SIRE

GREAT GRAND SIRE

GREAT GRAND DAM

GRAND DAM

GREAT GRAND SIRE

GREAT GRAND DAM

MEDICAL INFORMATION

INJURY OR ILLNESS

DATE	DESCRIPTION OR NATURE OF ILLNESS	TREATMENT

PARASITE CONTROL

DATE	METHOD OR DEWORMER	DATE	METHOD OR DEWORMER

TESTING RECORD

DATE	TEST PERFORMED (CAE, CL, TB...)	RESULT	DATE	TEST PERFORMED (CAE, CL, TB...)	RESULT

INJURY OR ILLNESS

DATE	TARGET DISEASE	DRUG OR SUPPLEMENT USED	DOSAGE	RESULTS

DOE'S KIDDING RECORD

DOE'S NAME:	

DATE BREED	KIDDING DATE	# OF KIDS	SEX D/B	NAME OF KID	SIRE OF KID	WEIGHT	TATTOO

BUCK'S RECORD OF PROGENY

DOE'S NAME:	

YEAR	BRED TO	KIDS	DOE/BUCK

GOAT RECORD

GOAT'S NAME:		IDENTIFICATION:	
BREED:	DATE OF BIRTH:		DATE OF WEANED:

WEIGHT (POUNDS)

BIRTH	JAN	FEB	MAR	APR	MAY	JUN	JUL	AUG	SEPT	OCT	NOV	DEC	FINAL

FEED RECORD

	JAN	FEB	MAR	APR	MAY	JUN	JUL	AUG	SEPT	OCT	NOV	DEC	TOTAL
GRAIN													
GRAIN													
PASTURE													

MILK PRODUCTION

GOAT'S NAME:		IDENTIFICATION:	
BREED:	DATE OF BIRTH:	KIDDING DATE:	

JANUARY		AVERAGE LBS / DAY X 31 DAYS =		LBS
FEBRUARY		AVERAGE LBS / DAY X 31 DAYS =		LBS
MARCH		AVERAGE LBS / DAY X 31 DAYS =		LBS
APRIL		AVERAGE LBS / DAY X 31 DAYS =		LBS
MAY		AVERAGE LBS / DAY X 31 DAYS =		LBS
JUNE		AVERAGE LBS / DAY X 31 DAYS =		LBS
JULY		AVERAGE LBS / DAY X 31 DAYS =		LBS
AUGUST		AVERAGE LBS / DAY X 31 DAYS =		LBS
SEPTEMBER		AVERAGE LBS / DAY X 31 DAYS =		LBS
OCTOBER		AVERAGE LBS / DAY X 31 DAYS =		LBS
NOVEMBER		AVERAGE LBS / DAY X 31 DAYS =		LBS
DECEMBER		AVERAGE LBS / DAY X 31 DAYS =		LBS
YEARLY TOTAL MILK PRODUCED =				LBS

TOTAL VALUE OF MILK PRODUCED FOR THE YEAR

	LBS X $		VALUE PER LBS =	

GOAT INFORMATION

PHOTO

NAME	☐ BUCK	☐ DOE
BREED	BIRTH DATE:	
DATE ACQUIRED:	HOW ACQUIRED: ☐ BORN ON FARM ☐ PURCHASED ☐ LEASED	
COLORS / IDENTIFYING MARKS:		
PURPOSE: ☐ MILK ☐ MEAT ☐ PET ☐ OTHER		

PEDIGREE CHART

SIRE

GRAND SIRE

GREAT GRAND SIRE

GREAT GRAND DAM

GRAND DAM

GREAT GRAND SIRE

GREAT GRAND DAM

DAM

GRAND SIRE

GREAT GRAND SIRE

GREAT GRAND DAM

GRAND DAM

GREAT GRAND SIRE

GREAT GRAND DAM

MEDICAL INFORMATION

INJURY OR ILLNESS

DATE	DESCRIPTION OR NATURE OF ILLNESS	TREATMENT

PARASITE CONTROL

DATE	METHOD OR DEWORMER	DATE	METHOD OR DEWORMER

TESTING RECORD

DATE	TEST PERFORMED (CAE, CL, TB...)	RESULT	DATE	TEST PERFORMED (CAE, CL, TB...)	RESULT

INJURY OR ILLNESS

DATE	TARGET DISEASE	DRUG OR SUPPLEMENT USED	DOSAGE	RESULTS

DOE'S KIDDING RECORD

DOE'S NAME:

DATE BREED	KIDDING DATE	# OF KIDS	SEX D/B	NAME OF KID	SIRE OF KID	WEIGHT	TATTOO

BUCK'S RECORD OF PROGENY

DOE'S NAME:	

YEAR	BRED TO	KIDS	DOE/BUCK

GOAT RECORD

GOAT'S NAME:		IDENTIFICATION:	
BREED:	DATE OF BIRTH:		DATE OF WEANED:

WEIGHT (POUNDS)

BIRTH	JAN	FEB	MAR	APR	MAY	JUN	JUL	AUG	SEPT	OCT	NOV	DEC	FINAL

FEED RECORD

	JAN	FEB	MAR	APR	MAY	JUN	JUL	AUG	SEPT	OCT	NOV	DEC	TOTAL
GRAIN													
GRAIN													
PASTURE													

MILK PRODUCTION

GOAT'S NAME:		IDENTIFICATION:	
BREED:	DATE OF BIRTH:	KIDDING DATE:	

JANUARY		AVERAGE LBS / DAY X 31 DAYS =		LBS
FEBRUARY		AVERAGE LBS / DAY X 31 DAYS =		LBS
MARCH		AVERAGE LBS / DAY X 31 DAYS =		LBS
APRIL		AVERAGE LBS / DAY X 31 DAYS =		LBS
MAY		AVERAGE LBS / DAY X 31 DAYS =		LBS
JUNE		AVERAGE LBS / DAY X 31 DAYS =		LBS
JULY		AVERAGE LBS / DAY X 31 DAYS =		LBS
AUGUST		AVERAGE LBS / DAY X 31 DAYS =		LBS
SEPTEMBER		AVERAGE LBS / DAY X 31 DAYS =		LBS
OCTOBER		AVERAGE LBS / DAY X 31 DAYS =		LBS
NOVEMBER		AVERAGE LBS / DAY X 31 DAYS =		LBS
DECEMBER		AVERAGE LBS / DAY X 31 DAYS =		LBS
YEARLY TOTAL MILK PRODUCED =				LBS

TOTAL VALUE OF MILK PRODUCED FOR THE YEAR

	LBS X $		VALUE PER LBS =	

GOAT INFORMATION

PHOTO

NAME	☐ BUCK	☐ DOE
BREED	BIRTH DATE:	

DATE ACQUIRED:	HOW ACQUIRED: ☐ BORN ON FARM ☐ PURCHASED ☐ LEASED

COLORS / IDENTIFYING MARKS:

PURPOSE:	☐ MILK	☐ MEAT	☐ PET	☐ OTHER

PEDIGREE CHART

SIRE

- GRAND SIRE
  - GREAT GRAND SIRE
  - GREAT GRAND DAM
- GRAND DAM
  - GREAT GRAND SIRE
  - GREAT GRAND DAM

DAM

- GRAND SIRE
  - GREAT GRAND SIRE
  - GREAT GRAND DAM
- GRAND DAM
  - GREAT GRAND SIRE
  - GREAT GRAND DAM

MEDICAL INFORMATION

INJURY OR ILLNESS

DATE	DESCRIPTION OR NATURE OF ILLNESS	TREATMENT

PARASITE CONTROL

DATE	METHOD OR DEWORMER	DATE	METHOD OR DEWORMER

TESTING RECORD

DATE	TEST PERFORMED (CAE, CL, TB...)	RESULT	DATE	TEST PERFORMED (CAE, CL, TB...)	RESULT

INJURY OR ILLNESS

DATE	TARGET DISEASE	DRUG OR SUPPLEMENT USED	DOSAGE	RESULTS

DOE'S KIDDING RECORD

DOE'S NAME:

DATE BREED	KIDDING DATE	# OF KIDS	SEX D/B	NAME OF KID	SIRE OF KID	WEIGHT	TATTOO

BUCK'S RECORD OF PROGENY

DOE'S NAME:	

YEAR	BRED TO	KIDS	DOE/BUCK

GOAT RECORD

GOAT'S NAME:		IDENTIFICATION:	
BREED:	DATE OF BIRTH:	DATE OF WEANED:	

WEIGHT (POUNDS)

BIRTH	JAN	FEB	MAR	APR	MAY	JUN	JUL	AUG	SEPT	OCT	NOV	DEC	FINAL

FEED RECORD

	JAN	FEB	MAR	APR	MAY	JUN	JUL	AUG	SEPT	OCT	NOV	DEC	TOTAL
GRAIN													
GRAIN													
PASTURE													

MILK PRODUCTION

GOAT'S NAME:		IDENTIFICATION:		
BREED:	DATE OF BIRTH:		KIDDING DATE:	

JANUARY		AVERAGE LBS / DAY X 31 DAYS =		LBS
FEBRUARY		AVERAGE LBS / DAY X 31 DAYS =		LBS
MARCH		AVERAGE LBS / DAY X 31 DAYS =		LBS
APRIL		AVERAGE LBS / DAY X 31 DAYS =		LBS
MAY		AVERAGE LBS / DAY X 31 DAYS =		LBS
JUNE		AVERAGE LBS / DAY X 31 DAYS =		LBS
JULY		AVERAGE LBS / DAY X 31 DAYS =		LBS
AUGUST		AVERAGE LBS / DAY X 31 DAYS =		LBS
SEPTEMBER		AVERAGE LBS / DAY X 31 DAYS =		LBS
OCTOBER		AVERAGE LBS / DAY X 31 DAYS =		LBS
NOVEMBER		AVERAGE LBS / DAY X 31 DAYS =		LBS
DECEMBER		AVERAGE LBS / DAY X 31 DAYS =		LBS
YEARLY TOTAL MILK PRODUCED =				LBS

TOTAL VALUE OF MILK PRODUCED FOR THE YEAR

	LBS X $		VALUE PER LBS =	

GOAT INFORMATION

PHOTO

NAME	☐ BUCK	☐ DOE
BREED	BIRTH DATE:	
DATE ACQUIRED:	HOW ACQUIRED: ☐ BORN ON FARM ☐ PURCHASED ☐ LEASED	
COLORS / IDENTIFYING MARKS:		
PURPOSE: ☐ MILK ☐ MEAT ☐ PET ☐ OTHER		

PEDIGREE CHART

- SIRE
  - GRAND SIRE
    - GREAT GRAND SIRE
    - GREAT GRAND DAM
  - GRAND DAM
    - GREAT GRAND SIRE
    - GREAT GRAND DAM
- DAM
  - GRAND SIRE
    - GREAT GRAND SIRE
    - GREAT GRAND DAM
  - GRAND DAM
    - GREAT GRAND SIRE
    - GREAT GRAND DAM

MEDICAL INFORMATION

INJURY OR ILLNESS

DATE	DESCRIPTION OR NATURE OF ILLNESS	TREATMENT

PARASITE CONTROL

DATE	METHOD OR DEWORMER	DATE	METHOD OR DEWORMER

TESTING RECORD

DATE	TEST PERFORMED (CAE, CL, TB...)	RESULT	DATE	TEST PERFORMED (CAE, CL, TB...)	RESULT

INJURY OR ILLNESS

DATE	TARGET DISEASE	DRUG OR SUPPLEMENT USED	DOSAGE	RESULTS

DOE'S KIDDING RECORD

DOE'S NAME:

DATE BREED	KIDDING DATE	# OF KIDS	SEX D/B	NAME OF KID	SIRE OF KID	WEIGHT	TATTOO

BUCK'S RECORD OF PROGENY

DOE'S NAME:	

YEAR	BRED TO	KIDS	DOE/BUCK

GOAT RECORD

GOAT'S NAME:		IDENTIFICATION:	
BREED:	DATE OF BIRTH:		DATE OF WEANED:

WEIGHT (POUNDS)

BIRTH	JAN	FEB	MAR	APR	MAY	JUN	JUL	AUG	SEPT	OCT	NOV	DEC	FINAL

FEED RECORD

	JAN	FEB	MAR	APR	MAY	JUN	JUL	AUG	SEPT	OCT	NOV	DEC	TOTAL
GRAIN													
GRAIN													
PASTURE													

MILK PRODUCTION

GOAT'S NAME:		IDENTIFICATION:	
BREED:	DATE OF BIRTH:	KIDDING DATE:	

JANUARY		AVERAGE LBS / DAY X 31 DAYS =		LBS
FEBRUARY		AVERAGE LBS / DAY X 31 DAYS =		LBS
MARCH		AVERAGE LBS / DAY X 31 DAYS =		LBS
APRIL		AVERAGE LBS / DAY X 31 DAYS =		LBS
MAY		AVERAGE LBS / DAY X 31 DAYS =		LBS
JUNE		AVERAGE LBS / DAY X 31 DAYS =		LBS
JULY		AVERAGE LBS / DAY X 31 DAYS =		LBS
AUGUST		AVERAGE LBS / DAY X 31 DAYS =		LBS
SEPTEMBER		AVERAGE LBS / DAY X 31 DAYS =		LBS
OCTOBER		AVERAGE LBS / DAY X 31 DAYS =		LBS
NOVEMBER		AVERAGE LBS / DAY X 31 DAYS =		LBS
DECEMBER		AVERAGE LBS / DAY X 31 DAYS =		LBS
YEARLY TOTAL MILK PRODUCED =				LBS

TOTAL VALUE OF MILK PRODUCED FOR THE YEAR

	LBS X $		VALUE PER LBS =	

GOAT INFORMATION

PHOTO

NAME	☐ BUCK	☐ DOE
BREED	BIRTH DATE:	

DATE ACQUIRED:	HOW ACQUIRED: ☐ BORN ON FARM ☐ PURCHASED ☐ LEASED

COLORS / IDENTIFYING MARKS:

PURPOSE:	☐ MILK	☐ MEAT	☐ PET	☐ OTHER

PEDIGREE CHART

- SIRE
  - GRAND SIRE
    - GREAT GRAND SIRE
    - GREAT GRAND DAM
  - GRAND DAM
    - GREAT GRAND SIRE
    - GREAT GRAND DAM
- DAM
  - GRAND SIRE
    - GREAT GRAND SIRE
    - GREAT GRAND DAM
  - GRAND DAM
    - GREAT GRAND SIRE
    - GREAT GRAND DAM

MEDICAL INFORMATION

INJURY OR ILLNESS

DATE	DESCRIPTION OR NATURE OF ILLNESS	TREATMENT

PARASITE CONTROL

DATE	METHOD OR DEWORMER	DATE	METHOD OR DEWORMER

TESTING RECORD

DATE	TEST PERFORMED (CAE, CL, TB...)	RESULT	DATE	TEST PERFORMED (CAE, CL, TB...)	RESULT

INJURY OR ILLNESS

DATE	TARGET DISEASE	DRUG OR SUPPLEMENT USED	DOSAGE	RESULTS

DOE'S KIDDING RECORD

DOE'S NAME:

DATE BREED	KIDDING DATE	# OF KIDS	SEX D/B	NAME OF KID	SIRE OF KID	WEIGHT	TATTOO

BUCK'S RECORD OF PROGENY

DOE'S NAME:

YEAR	BRED TO	KIDS	DOE/BUCK

GOAT RECORD

GOAT'S NAME:		IDENTIFICATION:	
BREED:	DATE OF BIRTH:	DATE OF WEANED:	

WEIGHT (POUNDS)

BIRTH	JAN	FEB	MAR	APR	MAY	JUN	JUL	AUG	SEPT	OCT	NOV	DEC	FINAL

FEED RECORD

	JAN	FEB	MAR	APR	MAY	JUN	JUL	AUG	SEPT	OCT	NOV	DEC	TOTAL
GRAIN													
GRAIN													
PASTURE													

MILK PRODUCTION

GOAT'S NAME:		IDENTIFICATION:	
BREED:	DATE OF BIRTH:	KIDDING DATE:	

JANUARY		AVERAGE LBS / DAY X 31 DAYS =		LBS
FEBRUARY		AVERAGE LBS / DAY X 31 DAYS =		LBS
MARCH		AVERAGE LBS / DAY X 31 DAYS =		LBS
APRIL		AVERAGE LBS / DAY X 31 DAYS =		LBS
MAY		AVERAGE LBS / DAY X 31 DAYS =		LBS
JUNE		AVERAGE LBS / DAY X 31 DAYS =		LBS
JULY		AVERAGE LBS / DAY X 31 DAYS =		LBS
AUGUST		AVERAGE LBS / DAY X 31 DAYS =		LBS
SEPTEMBER		AVERAGE LBS / DAY X 31 DAYS =		LBS
OCTOBER		AVERAGE LBS / DAY X 31 DAYS =		LBS
NOVEMBER		AVERAGE LBS / DAY X 31 DAYS =		LBS
DECEMBER		AVERAGE LBS / DAY X 31 DAYS =		LBS
YEARLY TOTAL MILK PRODUCED =				LBS

TOTAL VALUE OF MILK PRODUCED FOR THE YEAR

	LBS X $		VALUE PER LBS =	

GOAT INFORMATION

PHOTO

NAME		☐ BUCK	☐ DOE
BREED		BIRTH DATE:	
DATE ACQUIRED:	HOW ACQUIRED: ☐ BORN ON FARM ☐ PURCHASED ☐ LEASED		
COLORS / IDENTIFYING MARKS:			
PURPOSE: ☐ MILK ☐ MEAT ☐ PET	☐ OTHER		

PEDIGREE CHART

- SIRE
  - GRAND SIRE
    - GREAT GRAND SIRE
    - GREAT GRAND DAM
  - GRAND DAM
    - GREAT GRAND SIRE
    - GREAT GRAND DAM
- DAM
  - GRAND SIRE
    - GREAT GRAND SIRE
    - GREAT GRAND DAM
  - GRAND DAM
    - GREAT GRAND SIRE
    - GREAT GRAND DAM

MEDICAL INFORMATION

INJURY OR ILLNESS

DATE	DESCRIPTION OR NATURE OF ILLNESS	TREATMENT

PARASITE CONTROL

DATE	METHOD OR DEWORMER	DATE	METHOD OR DEWORMER

TESTING RECORD

DATE	TEST PERFORMED (CAE, CL, TB...)	RESULT	DATE	TEST PERFORMED (CAE, CL, TB...)	RESULT

INJURY OR ILLNESS

DATE	TARGET DISEASE	DRUG OR SUPPLEMENT USED	DOSAGE	RESULTS

DOE'S KIDDING RECORD

DOE'S NAME:

DATE BREED	KIDDING DATE	# OF KIDS	SEX D/B	NAME OF KID	SIRE OF KID	WEIGHT	TATTOO

BUCK'S RECORD OF PROGENY

DOE'S NAME:	

YEAR	BRED TO	KIDS	DOE/BUCK

GOAT RECORD

GOAT'S NAME:		IDENTIFICATION:	
BREED:	DATE OF BIRTH:	DATE OF WEANED:	

WEIGHT (POUNDS)

BIRTH	JAN	FEB	MAR	APR	MAY	JUN	JUL	AUG	SEPT	OCT	NOV	DEC	FINAL

FEED RECORD

	JAN	FEB	MAR	APR	MAY	JUN	JUL	AUG	SEPT	OCT	NOV	DEC	TOTAL
GRAIN													
GRAIN													
PASTURE													

MILK PRODUCTION

GOAT'S NAME:		IDENTIFICATION:	
BREED:	DATE OF BIRTH:	KIDDING DATE:	

JANUARY		AVERAGE LBS / DAY X 31 DAYS =		LBS
FEBRUARY		AVERAGE LBS / DAY X 31 DAYS =		LBS
MARCH		AVERAGE LBS / DAY X 31 DAYS =		LBS
APRIL		AVERAGE LBS / DAY X 31 DAYS =		LBS
MAY		AVERAGE LBS / DAY X 31 DAYS =		LBS
JUNE		AVERAGE LBS / DAY X 31 DAYS =		LBS
JULY		AVERAGE LBS / DAY X 31 DAYS =		LBS
AUGUST		AVERAGE LBS / DAY X 31 DAYS =		LBS
SEPTEMBER		AVERAGE LBS / DAY X 31 DAYS =		LBS
OCTOBER		AVERAGE LBS / DAY X 31 DAYS =		LBS
NOVEMBER		AVERAGE LBS / DAY X 31 DAYS =		LBS
DECEMBER		AVERAGE LBS / DAY X 31 DAYS =		LBS
YEARLY TOTAL MILK PRODUCED =				LBS

TOTAL VALUE OF MILK PRODUCED FOR THE YEAR

	LBS X $		VALUE PER LBS =	

GOAT INFORMATION

PHOTO

NAME		☐ BUCK	☐ DOE
BREED		BIRTH DATE:	
DATE ACQUIRED:	HOW ACQUIRED: ☐ BORN ON FARM ☐ PURCHASED ☐ LEASED		
COLORS / IDENTIFYING MARKS:			
PURPOSE: ☐ MILK ☐ MEAT ☐ PET ☐ OTHER			

PEDIGREE CHART

SIRE

GRAND SIRE

GREAT GRAND SIRE

GREAT GRAND DAM

GRAND DAM

GREAT GRAND SIRE

GREAT GRAND DAM

DAM

GRAND SIRE

GREAT GRAND SIRE

GREAT GRAND DAM

GRAND DAM

GREAT GRAND SIRE

GREAT GRAND DAM

MEDICAL INFORMATION

INJURY OR ILLNESS

DATE	DESCRIPTION OR NATURE OF ILLNESS	TREATMENT

PARASITE CONTROL

DATE	METHOD OR DEWORMER	DATE	METHOD OR DEWORMER

TESTING RECORD

DATE	TEST PERFORMED (CAE, CL, TB...)	RESULT	DATE	TEST PERFORMED (CAE, CL, TB...)	RESULT

INJURY OR ILLNESS

DATE	TARGET DISEASE	DRUG OR SUPPLEMENT USED	DOSAGE	RESULTS

DOE'S KIDDING RECORD

DOE'S NAME:	

DATE BREED	KIDDING DATE	# OF KIDS	SEX D/B	NAME OF KID	SIRE OF KID	WEIGHT	TATTOO

BUCK'S RECORD OF PROGENY

DOE'S NAME:	

YEAR	BRED TO	KIDS	DOE/BUCK

GOAT RECORD

GOAT'S NAME:		IDENTIFICATION:
BREED:	DATE OF BIRTH:	DATE OF WEANED:

WEIGHT (POUNDS)

BIRTH	JAN	FEB	MAR	APR	MAY	JUN	JUL	AUG	SEPT	OCT	NOV	DEC	FINAL

FEED RECORD

	JAN	FEB	MAR	APR	MAY	JUN	JUL	AUG	SEPT	OCT	NOV	DEC	TOTAL
GRAIN													
GRAIN													
PASTURE													

MILK PRODUCTION

GOAT'S NAME:		IDENTIFICATION:	
BREED:	DATE OF BIRTH:	KIDDING DATE:	

JANUARY		AVERAGE LBS / DAY X 31 DAYS =		LBS
FEBRUARY		AVERAGE LBS / DAY X 31 DAYS =		LBS
MARCH		AVERAGE LBS / DAY X 31 DAYS =		LBS
APRIL		AVERAGE LBS / DAY X 31 DAYS =		LBS
MAY		AVERAGE LBS / DAY X 31 DAYS =		LBS
JUNE		AVERAGE LBS / DAY X 31 DAYS =		LBS
JULY		AVERAGE LBS / DAY X 31 DAYS =		LBS
AUGUST		AVERAGE LBS / DAY X 31 DAYS =		LBS
SEPTEMBER		AVERAGE LBS / DAY X 31 DAYS =		LBS
OCTOBER		AVERAGE LBS / DAY X 31 DAYS =		LBS
NOVEMBER		AVERAGE LBS / DAY X 31 DAYS =		LBS
DECEMBER		AVERAGE LBS / DAY X 31 DAYS =		LBS
YEARLY TOTAL MILK PRODUCED =				LBS
TOTAL VALUE OF MILK PRODUCED FOR THE YEAR				
	LBS X $		VALUE PER LBS =	

GOAT INFORMATION

PHOTO

NAME		☐ BUCK	☐ DOE
BREED		BIRTH DATE:	
DATE ACQUIRED:	HOW ACQUIRED: ☐ BORN ON FARM ☐ PURCHASED ☐ LEASED		
COLORS / IDENTIFYING MARKS:			
PURPOSE: ☐ MILK ☐ MEAT ☐ PET ☐ OTHER			

PEDIGREE CHART

SIRE

GRAND SIRE

GREAT GRAND SIRE

GREAT GRAND DAM

GRAND DAM

GREAT GRAND SIRE

GREAT GRAND DAM

DAM

GRAND SIRE

GREAT GRAND SIRE

GREAT GRAND DAM

GRAND DAM

GREAT GRAND SIRE

GREAT GRAND DAM

MEDICAL INFORMATION

INJURY OR ILLNESS

DATE	DESCRIPTION OR NATURE OF ILLNESS	TREATMENT

PARASITE CONTROL

DATE	METHOD OR DEWORMER	DATE	METHOD OR DEWORMER

TESTING RECORD

DATE	TEST PERFORMED (CAE, CL, TB...)	RESULT	DATE	TEST PERFORMED (CAE, CL, TB...)	RESULT

INJURY OR ILLNESS

DATE	TARGET DISEASE	DRUG OR SUPPLEMENT USED	DOSAGE	RESULTS

DOE'S KIDDING RECORD

DOE'S NAME:	

DATE BREED	KIDDING DATE	# OF KIDS	SEX D/B	NAME OF KID	SIRE OF KID	WEIGHT	TATTOO

BUCK'S RECORD OF PROGENY

DOE'S NAME:	

YEAR	BRED TO	KIDS	DOE/BUCK

GOAT RECORD

GOAT'S NAME:		IDENTIFICATION:	
BREED:	**DATE OF BIRTH:**		**DATE OF WEANED:**

WEIGHT (POUNDS)

BIRTH	JAN	FEB	MAR	APR	MAY	JUN	JUL	AUG	SEPT	OCT	NOV	DEC	FINAL

FEED RECORD

	JAN	FEB	MAR	APR	MAY	JUN	JUL	AUG	SEPT	OCT	NOV	DEC	TOTAL
GRAIN													
GRAIN													
PASTURE													

MILK PRODUCTION

GOAT'S NAME:		IDENTIFICATION:		
BREED:	DATE OF BIRTH:	KIDDING DATE:		

JANUARY		AVERAGE LBS / DAY X 31 DAYS =		LBS
FEBRUARY		AVERAGE LBS / DAY X 31 DAYS =		LBS
MARCH		AVERAGE LBS / DAY X 31 DAYS =		LBS
APRIL		AVERAGE LBS / DAY X 31 DAYS =		LBS
MAY		AVERAGE LBS / DAY X 31 DAYS =		LBS
JUNE		AVERAGE LBS / DAY X 31 DAYS =		LBS
JULY		AVERAGE LBS / DAY X 31 DAYS =		LBS
AUGUST		AVERAGE LBS / DAY X 31 DAYS =		LBS
SEPTEMBER		AVERAGE LBS / DAY X 31 DAYS =		LBS
OCTOBER		AVERAGE LBS / DAY X 31 DAYS =		LBS
NOVEMBER		AVERAGE LBS / DAY X 31 DAYS =		LBS
DECEMBER		AVERAGE LBS / DAY X 31 DAYS =		LBS
YEARLY TOTAL MILK PRODUCED =				LBS
TOTAL VALUE OF MILK PRODUCED FOR THE YEAR				
	LBS X $		VALUE PER LBS =	

NOTES